Health
72

细菌汤
A Soup of Bacteria

Gunter Pauli

冈特·鲍利 著

李欢欢 译

www.xuelinpress.com

丛书编委会

主　任：贾　峰

副主任：何家振　郑立明

委　员：牛玲娟　李原原　李曙东　吴建民　彭　勇
　　　　冯　缨　靳增江

丛书出版委员会

主　任：段学俭

副主任：匡志强　张　蓉

成　员：叶　刚　李晓梅　魏　来　徐雅清　田振军
　　　　蔡雩奇

特别感谢以下热心人士对译稿润色工作的支持：

姜竹青　韩　笑　杨　爽　周依奇　于　哲　阳平坚
李雪红　汪　楠　单　威　查振旺　李海红　姚爱静
朱　国　彭　江　于洪英　隋淑光　严　岷

目录

细菌汤	4
你知道吗?	22
想一想	26
自己动手!	27
学科知识	28
情感智慧	29
艺术	29
思维拓展	30
动手能力	30
故事灵感来自	31

Contents

A Soup of Bacteria	4
Did you know?	22
Think about it	26
Do it yourself!	27
Academic Knowledge	28
Emotional Intelligence	29
The Arts	29
Systems: Making the Connections	30
Capacity to Implement	30
This fable is inspired by	31

一条金枪鱼在塔斯曼海游来游去,他发现了一小片红藻林。

"你有和我一样的颜色。"金枪鱼说出了自己的看法。

A tuna fish swirls around the Tasman Sea and finds a small forest of red seaweed.

"You are the same colour as me," observes the tuna.

……一小片红藻林……

...a small forest of red seaweed...

我的名字叫美味

My name is delicious

"是呀，那你知道我的名字叫美味吗？"红藻回答道。

"哦，那咱俩都很美味，"金枪鱼边说边钻进了海藻丛，"不过你好粗糙呀。你一点都不柔软。"

"Yes, and did you know that my name is delicious?" responds the seaweed.

"Oh, so we both are delicious," remarks the tuna as he swims through the weeds. "But you feel so coarse. You're not soft at all."

"啊，那是因为我是无菌的呀。"

"世界上根本就不存在无菌的东西！地球上的生命起源于海洋，而最早出现的物种就是细菌。细菌王国是自然界里最大的！"

"Ah, that is because I am bacteria-free."

"There's no such thing in the world! Life on earth emerged from the sea, and the first living species were bacteria. It's the biggest kingdom in nature!"

我是无菌的

I am bacteria-free

一碗细菌汤……

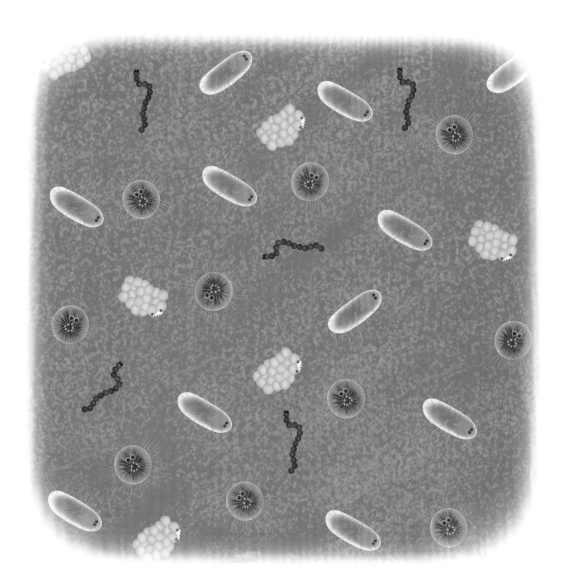

A soup of bacteria…

"那还用说嘛——海洋简直就像一碗细菌汤！"

"既然你知道你就生活在细菌汤里，你怎么能说自己是无菌的呢？"

"因为我把细菌变成了聋子。"

"Tell me about it – the sea is like a soup of bacteria!"

"So if you know you're living in a bacteria soup, how can you claim to be free of them?"

"Because I make them deaf."

"哦，别胡扯了!我听说过鲸会交流，植物会唱歌，可你居然告诉我你能叫得如此尖锐，以致于细菌们都无法听见彼此的声音了？"

"不，我不是尖叫——我只是塞住了他们的耳朵。"红藻说。

"细菌没有耳朵，因此你不可能塞住他们的耳朵。"金枪鱼坚持道。

"Oh, come on! I've heard about whales communicating and plants singing, but you're telling me you can scream so loud that the bacteria can't hear each other?"

"No, I'm not screaming – I'm plugging their ears," says the seaweed.

"Bacteria don't have ears, so you can't plug them," insists the tuna.

······我听说过······

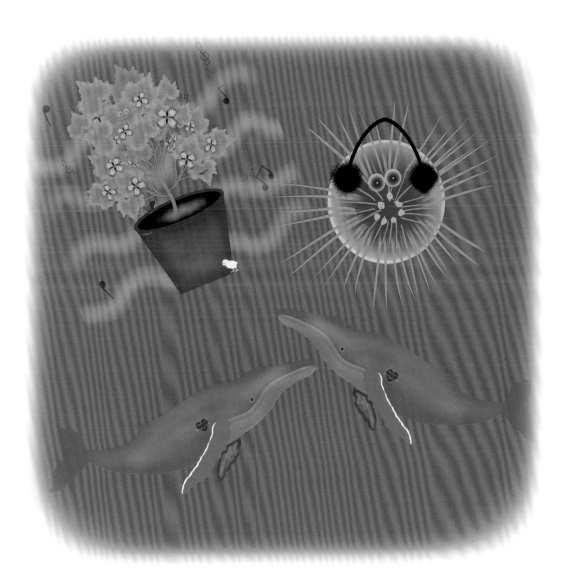

...I've heard about...

它们叫作受体，不叫耳朵

They're called receptors, not ears

"呃，它们叫作受体，不叫耳朵，不过这是一回事。我释放这些微小的化学物质，塞住了他们的耳朵。"

"那你为什么要那么做呢？"

"Well, they're called receptors, not ears, but it's the same thing. I release these small chemicals that block their ears."

"And why do you do that?"

"嗯，"红藻解释道，"细菌无处不在，侵入一切生物，包括我。它有时让我们难以生存。"

"你有没有尝试过杀死他们？"金枪鱼好奇地问。

"杀死细菌？这种事还是留给人类吧——他们一直在尝试消灭细菌，可结果却是在慢慢杀死自己。他们永远不会成功的。"

"Well," says the seaweed, "bacteria are everywhere and invade everything, including me. It sometimes makes life impossible."

"Have you ever tried to kill them?" wonders the tuna.

"Killing bacteria? Leave that to humans – they try all the time but are slowly killing themselves. They'll never succeed."

细菌喜欢群居

Bacteria love to cuddle together

"因此你决定塞住他们的耳朵。接下来怎么办呢？"

"细菌喜欢群居。细菌群会形成一层薄薄的软膜，但只有细菌家族壮大到足够的数量，才能控制他们所寄生的寄主。"

"So you decided to plug their ears. Then what?"

"Bacteria love to cuddle together. They make a thin, soft film, but they need a quorum in the family so that they can control their host."

"因此当你塞住了他们的耳朵,他们就不知道家在哪儿,只能到别的地方定居,这样你就安全了。"

"我们是不是很聪明?你也应该试试——这些细菌最喜欢享用鱼肉了,他们会破坏你的肉质。这样也许你闻着都不像鱼了!"

……这仅仅是开始!……

"So when you plug their ears, they have no clue where their family is and settle somewhere else, leaving you at peace."

"Aren't we clever? You should try it too – these bacteria love feasting on fish, and they spoil your meat. And then maybe you wouldn't smell like fish!"

… AND IT ONLY HAS JUST BEGUN!…

……这仅仅是开始!……

…AND IT HAS ONLY JUST BEGUN!…

单个细菌不是问题。当一群细菌共同作用形成菌膜时，才有可能控制他们所寄生的寄主。

A single bacterium is not a problem. It's when bacteria work together to create a biofilm that sometimes try to control their host.

比起消灭单个细菌，你需要使用功效可能要高 1000 倍的抗生素才能消灭一个已形成菌膜的细菌群落。

You need an antibiotic perhaps 1 000 times stronger to kill a colony of bacteria formed into a biofilm than you would for a single bacterium.

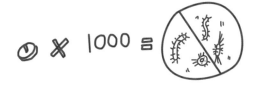

Bacteria communicate with each other using a chemical language. Bacteria not only talk, they also coordinate and undertake joint initiatives.

细菌使用化学语言进行沟通。细菌不仅相互交流，他们还相互协作，采取联合行动。

Bacteria need to have a minimum number of members to start coordinating. This is called "quorum sensing". They also sense how many "other" bacteria are present.

细菌成员必须达到一个最低数值，才能开始协作，这叫群体感应。他们还能感知到有多少其他种类的细菌存在。

The type and quantity of microorganisms in your gut can prevent or encourage disease, including mental health, making your gut function like a second brain.

人体肠道内的微生物种类和数量不同,有些能预防疾病,有些却会诱发疾病,甚至影响心理健康,这就使得肠道起到了人体第二大脑的作用。

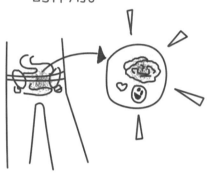

A swimmer swallowing a mouthful of seawater may be consuming more than 1 000 types of bacteria.

一名游泳者吞下一口海水,也许就喝掉了1000种细菌。

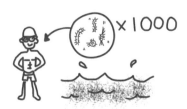

Pickled food, is fermented food, like miso, natto, tempeh, kimchee, kombucha, sauerkraut, pickles, olives and raw milk cheeses.

腌制食品是一种发酵食品，如味噌、纳豆、印尼豆豉、韩国泡菜、康普茶、德国酸菜、各式腌菜、腌橄榄和生奶酪。

Tears are the best natural antibacterial substance produced by a human being.

眼泪是人体产生的最佳天然抗菌物质。

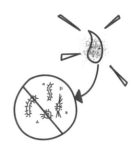

Think About It
想一想

Plants, animals and now bacteria seem to communicate with each other. So which language will you learn next?

植物、动物和细菌似乎都能相互交流。那么，你下一步打算学哪种语言呢？

你认为，你能单靠自己就获得成功，还是需要一个团体（至少有一定数量的朋友）才能让你立于不败之地？

Do you think you can win on your own, or do you need a quorum (a minimum number of friends) to make you invincible?

Will people ever win the war against bacteria and kill them all?

人类是否可以在对抗细菌的战争中获胜，并且消灭所有细菌？

如果你知道自己很聪明，那么炫耀自己的聪明是个好主意吗？

When you know when you are clever, is it a good idea to brag about it?

Try to get hold of a strong microscope. Take some water from the tap, and look for life in the sample. Then collect some saliva from your mouth, and look for life in there. Next, take some water that has been boiled, and look for life in there. Now leave the boiled water standing for 2 days. What do you see under the microscope? It seems impossible to get rid of bacteria, they are everywhere, so it's best to learn to live with them.

试着找来一台功能强大的显微镜。取一些自来水，在显微镜下寻找样本中的生命体。再从你嘴里取一些唾液，寻找其中的生命体。接下来，取一些凉开水，找找里面的生命体。现在把凉开水放置两天。你在显微镜下看到了什么？想摆脱细菌似乎很难，细菌无处不在，所以我们最好能学会与细菌共同生存。

TEACHER AND PARENT GUIDE

学科知识
Academic Knowledge

生物学	海洋红藻是一种红色的海草；细菌是地球上最原始的生命形式；细菌喜欢成群结队，以菌膜的形式，而不是自由浮动的细胞个体存在；超级病菌是对多种抗生素有抗性的细菌菌株；一升海水含有2万种不同的微生物；海洋中，90%生命体是微生物；细菌会在压力之下变异。
化 学	呋喃酮由海洋红藻制成，可以使细菌"失去听力"，这样不用杀死细菌就能避免细菌感染，还不会引起细菌的抗药性；肠道菌群对抗生素、加氯水、农药和抗菌皂非常敏感；眼泪中的溶解酶是人体最佳的抗菌抗病毒药剂，10分钟内能杀死90%～95%细菌；饮食中，发酵食品的种类越丰富越好，发酵食品可以增加肠道内微生物种类，既提供微量营养素，还有排毒作用；天然的杀菌剂包括氧气、碘酒、酒精和臭氧。
物 理	呋喃酮与细菌受体部位结合，会阻止细菌间的交流，但不会引起其他任何负作用。
工程学	维护石油和天然气管道包括去除管道内的菌膜；受到菌膜威胁的食品行业利用刺激性化学物质来控制细菌，尤其是奶制品、鱼类、家禽和即食呋喃酮行业。
经济学	呋喃酮可以应用于疾病治疗和医疗设施、受菌膜侵蚀的油气管道、空调、去污产品和污水处理方面。
伦理学	如果有其他选择不会引起细菌抗药性，我们怎么能用会引起细菌变异和产生抗体的化学物质呢？让生物变聋比杀死它要更好吗？改变比拒绝现状要好吗？
历 史	法国科学家于1844年最早记载了海洋红藻。
地 理	海洋红藻在塔斯曼海域大量生长，也见于太平洋沿岸地区；生物地理学研究显示没有相同的细菌汤，但在海洋的部分地区，有些细菌很常见，有些却很罕见。
数 学	数学模型可以预测传染病如何发展，如何预防传染病，何种公共卫生干预措施可以达到防制目的。
生活方式	世界上多个国家把发酵食物作为饮食的一部分：日本（味噌、印尼豆豉、纳豆、康普茶），韩国（泡菜），德国/法国（酸菜），比利时（酸乳酒）。
社会学	整个社会一味地想要消灭细菌，有着一套严谨的消灭细菌的策略，这样最终可能会导致人的免疫系统失效。
心理学	洁癖或细菌恐惧症是对污染物和细菌的病态恐惧，属于强迫性精神障碍。
系统论	滥用抗生素导致细菌变异，形成超级病毒；其实有很多天然的方法可供选择，如紫外线、臭氧和呋喃酮，这些都不会引起细菌变异。

教师与家长指南

情感智慧
Emotional Intelligence

金枪鱼　　金枪鱼善于社交，积极地开启话题，但他对海藻的回答有些自负。他对环境很敏感，并乐于分享他的观点和经历。金枪鱼不相信海藻说的话，并通过列举事实，来展示他的博学多才。金枪鱼有些自高自大，并开始质疑海藻，当他获得新信息时，还讥讽海藻。他否定海藻的解释，认为细菌没有耳朵，也没法堵上耳朵。他质疑海藻说的每一点，但最后终于明白了海藻的逻辑，确定细菌防制是可行的，而且他的话讲得好像早就知道这点似的。

海　藻　　海藻欢迎金枪鱼来闲聊，她很了解自己，毫无保留地分享自己的秘密。海藻并没有进行长篇大论的解释，而是简洁清晰地表述重点。当金枪鱼表示质疑时，她阐明了一下科学上的正确用词，指出让细菌耳聋而非杀死他们的原因。最后，明知听起来很自负，海藻仍表示她对自己的方法很满意。

艺术
The Arts

找出几张海洋红藻的图片：这些红色的分支相当漂亮。如果你住在太平洋沿岸，尤其是澳大利亚，你可以试着摘一片海洋红藻，不过要小心哦，这种海藻生长缓慢，你取下的任何一点红藻，都需要几年时间长回去。把红藻图片投影到银幕上，试着把红藻画下来。如何来显示这是无菌叶片呢？

TEACHER AND PARENT GUIDE

思维拓展
Systems: Making the Connections

人类已向细菌宣战：我们的唯一策略就是消灭细菌。然而，细菌分很多种，其中大部分是有益菌：没有它们无穷无尽的有效支持，我们可能没法存活。可惜，我们很少想到细菌独特的贡献或我们对其的依赖性。我们嘴里帮助消化的细菌数量比地球上的人口还多。人和细菌是共生关系，一个人身上约有100万亿个细菌细胞。细菌占人体10%的重量，人体内的细菌细胞数量是身体细胞的10倍。但是我们仍然把所有的细菌归类为病菌，并从很早开始，执着地想要消灭细菌。问题在于消灭所有的细菌，不可能不影响到我们自己的生活质量。我们已经介绍了那么多细菌防制措施，过量消耗抗生素或加氯水使我们变得虚弱：体内的有益菌受到化学物质的攻击，变异菌种和产生抗药性的菌种生存下来。近几年超级病毒的出现就是几十年来大量使用抗生素的后果。因此人类应该借鉴自然界的方法：盲目地消灭一切细菌，还不如阻止细菌彼此交流，从而使其不可能联合进攻其寄主。

动手能力
Capacity to Implement

我们需要控制有害的细菌，但我们不想杀死有益的细菌。列出所有不杀害微生物就能防制微生物的自然方式。我们想让大家理解：尽管抗生素是一种选择，但并非首选，它也许是治疗疾病的最后一种手段。勤洗手也许是最佳的预防措施。列出你的清单，确定不同的情况下，哪个是最好的选择。只有我们清楚地认识到如何展开无毒防制，我们才能维护地球上的生命免受细菌侵害。

教师与家长指南

故事灵感来自
皮特·斯坦伯格
Peter Steinberg

皮特·斯坦伯格曾在美国马里兰大学学习动物学，并获得加州大学博士学位。他在澳大利亚工作生活了 20 年，在跳水时注意到了海洋红藻的纹理。他的研究巅峰是一系列关于细菌群体效应的研究论文，还有几项专利。之后，他和新南威尔士大学的同事一起开创了生物信号公司。尽管公司没能把产品推向市场，但让人们对细菌防制有了新的看法。现在，皮特是悉尼海洋科学研究所的主任以及澳大利亚新南威尔士大学海洋港口项目负责人。

更多资讯

articles.mercola.com/sites/articles/archive/2012/12/29/probiotics-for-good-digestive-health.aspx

newsinhealth.nih.gov/issue/feb2014/feature1

www.abc.net.au/science/news/stories/s1702359.htm

http://www.biosignals.biostec.org/

图书在版编目（CIP）数据

细菌汤：汉英对照 /（比）鲍利著；李欢欢译. -- 上海：学林出版社，2015.6
（冈特生态童书. 第2辑）
ISBN 978-7-5486-0852-3

Ⅰ. ①细… Ⅱ. ①鲍… ②李… Ⅲ. ①生态环境－环境保护－儿童读物－汉、英 Ⅳ. ① X171.1-49

中国版本图书馆 CIP 数据核字（2015）第 086068 号

————————————————————————————

ⓒ 2015 Gunter Pauli
著作权合同登记号 图字 09-2015-446 号

冈特生态童书

细菌汤

作　　者——	冈特·鲍利
译　　者——	李欢欢
策　　划——	匡志强
责任编辑——	匡志强　蔡雩奇
装帧设计——	魏　来
出　　版——	上海世纪出版股份有限公司 学林出版社
	地　址：上海钦州南路81号　电话/传真：021-64515005
	网　址：www.xuelinpress.com
发　　行——	上海世纪出版股份有限公司发行中心
	（上海福建中路193号　网址：www.ewen.co）
印　　刷——	上海图宇印刷有限公司
开　　本——	710×1020　1/16
印　　张——	2
字　　数——	5万
版　　次——	2015年6月第1版
	2015年6月第1次印刷
书　　号——	ISBN 978-7-5486-0852-3/G·301
定　　价——	10.00元

（如发生印刷、装订质量问题，读者可向工厂调换）